Curious Creatures

NARWHALS

GAIL TERP

BLACK RABBIT BOOKS

Bolt is published by Black Rabbit Books
P.O. Box 227, Mankato, Minnesota, 56002
www.blackrabbitbooks.com

Marysa Storm, editor; Michael Sellner, designer and photo researcher

Library of Congress Cataloging-in-Publication Data
Names: Terp, Gail, 1951- author.
Title: Narwhals / by Gail Terp.
Description: Mankato, Minnesota : Black Rabbit Books, [2023] | Series: Bolt. Curious creatures | Includes bibliographical references and index. | Audience: Ages 8-12 | Audience: Grades 4-6 | Summary: "Narwhals are animals unlike any other, and curious and reluctant readers alike will be captivated by these exceptional creatures' life cycle, habitats, diet, and threats to survival through carefully leveled text, detailed infographics that emphasize visual literacy, and engaging, colorful photography"- Provided by publisher.
Identifiers: LCCN 2020016581 (print) | LCCN 2020016582 (ebook) | ISBN 9781623105662 (hardcover) | ISBN 9781644664919 (paperback) | ISBN 9781623105723 (ebook)
Subjects: LCSH: Narwhal–Juvenile literature.
Classification: LCC QL737.C433 T47 2023 (print) | LCC QL737.C433 (ebook) | DDC 599.5/43-dc23
LC record available at https://lccn.loc.gov/2020016581
LC ebook record available at https://lccn.loc.gov/2020016582

Image Credits

agefotostock: Russ Kinne, 18–19; Alamy: A & J Visage, 4–5; All Canada Photos, 1, 12–13; blickwinkel, 25; Blue Planet Archive: Bryan & Cherry Alexander, 26; Flickr: Our Breathing Planet, 16–17; Getty: VW Pics, 22–23; iStock: dottedhippo, 3, 6–7, 8–9, 18–19, 25, 29; Minden Pictures: David Shale, 10–11; Doug Allan, 11; Eric Baccega, 21; Espen Bergersen, 26; Flip Nicklin, 20, 29; Narwhal Tusk Research: Martin T. Nweeia, 29; Nat Geo Creative: Paul Nicklen, Cover; Science Source: Alex Rockheart, 14–15; Bryan & Cherry Alexander/, 6; Shutterstock: BearFotos, 25; Christian Musat, 25; Fujifilm123, 25; GooseFrol, 19

Contents

CHAPTER 1

A Narwhal in Action

It's spring and time for a female narwhal to **mate**. Four males come near. Each wants to prove it's the best. The males point their heads to the sky. Their long tusks stretch high above the water. Then they cross their tusks and whistle. It's not much of a fight. But still, the female chooses a winner.

How Big Is a Narwhal?

BODY
LENGTH
13 TO 20
FEET
(4 to 6 meters)

The Tusk

Narwhals are a type of whale. Their giant tusks make them different from other whales. Only narwhals have them. The tusk is just a really big tooth that grows through the upper lip. Most males have one tusk. Only a few have two. Scientists aren't completely sure what the tusks are for. Most agree males use them to get **mates**.

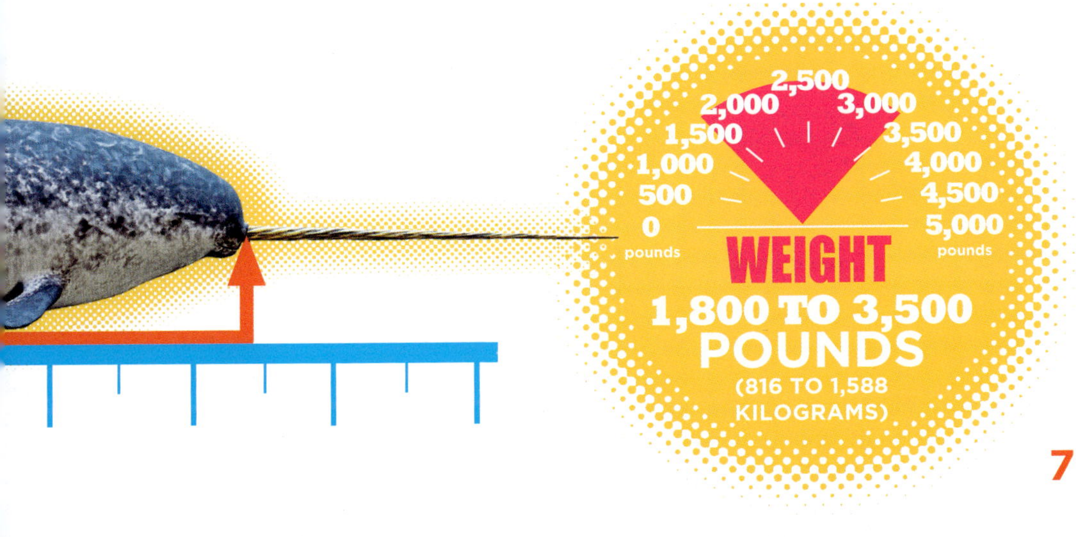

PARTS OF A NARWHAL

FLUKES

TUSK
(MALES ONLY)
EYES
FLIPPERS

CHAPTER 2

FOOD and Home

Narwhals eat different kinds of **prey**. They feed on several types of fish. They also eat shrimp and squid. Their prey lives in deep water. But catching prey is no problem for narwhals. They can dive deep into the ocean. When they find prey, they suck it in and swallow it whole.

Deep in the ocean, it's dark. To find prey, narwhals send out sounds. These sounds bounce off nearby prey. The narwhals listen. The bounced sounds tell them where the prey is.

An Icy, Watery Home

These whales are found in cold **arctic** seas. They mostly live at the edge of sea ice. Under their skin, narwhals have thick layers of **blubber**. It protects them from the cold.

Narwhals must be careful around the ice. It can quickly freeze. Pieces of ice can move together to close holes. Narwhals can get trapped. They can't get to the surface to breathe.

WHERE NARWHALS LIVE

Narwhal Range Map

Arctic Ocean
Pacific Ocean
Indian Ocean
Southern Ocean

CHAPTER 3

The Family

Narwhals are very **social** animals. They tend to live in groups of 15 to 20. In summer, groups often join together to create **herds** of thousands. These giant groups travel to areas with more ice to swim beneath.

narwhal
couple

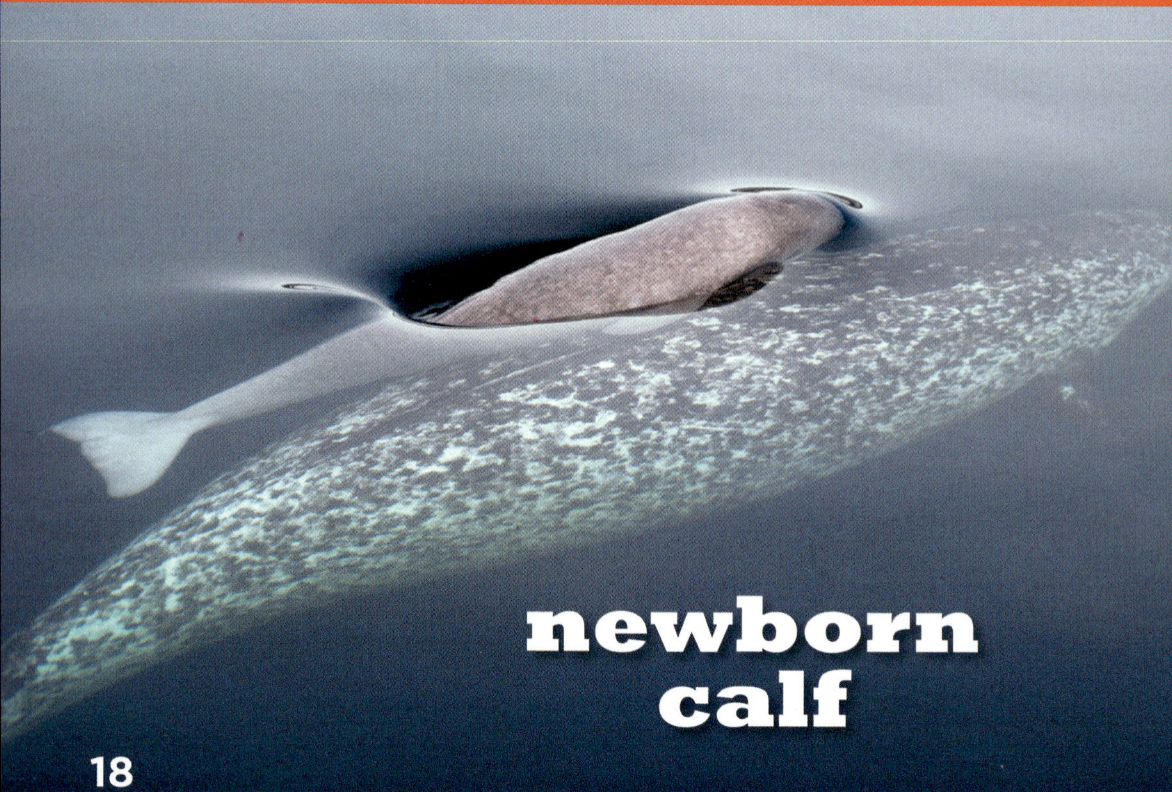
newborn
calf

Mating

Narwhals often mate in early spring. The following summer, females give birth to one calf. Newborns swim soon after they are born. Their tusks start growing within a year.

COMPARING LENGTHS

Growing Big

Young narwhals can't take care of themselves. Their mothers must nurse them for up to two years. Narwhal milk is very rich. This fatty milk helps build up calves' layers of blubber.

Some narwhals turn white when they grow old.

By the Numbers

4 to 7 YEARS

AGE WHEN FEMALE NARWHALS CAN HAVE BABIES

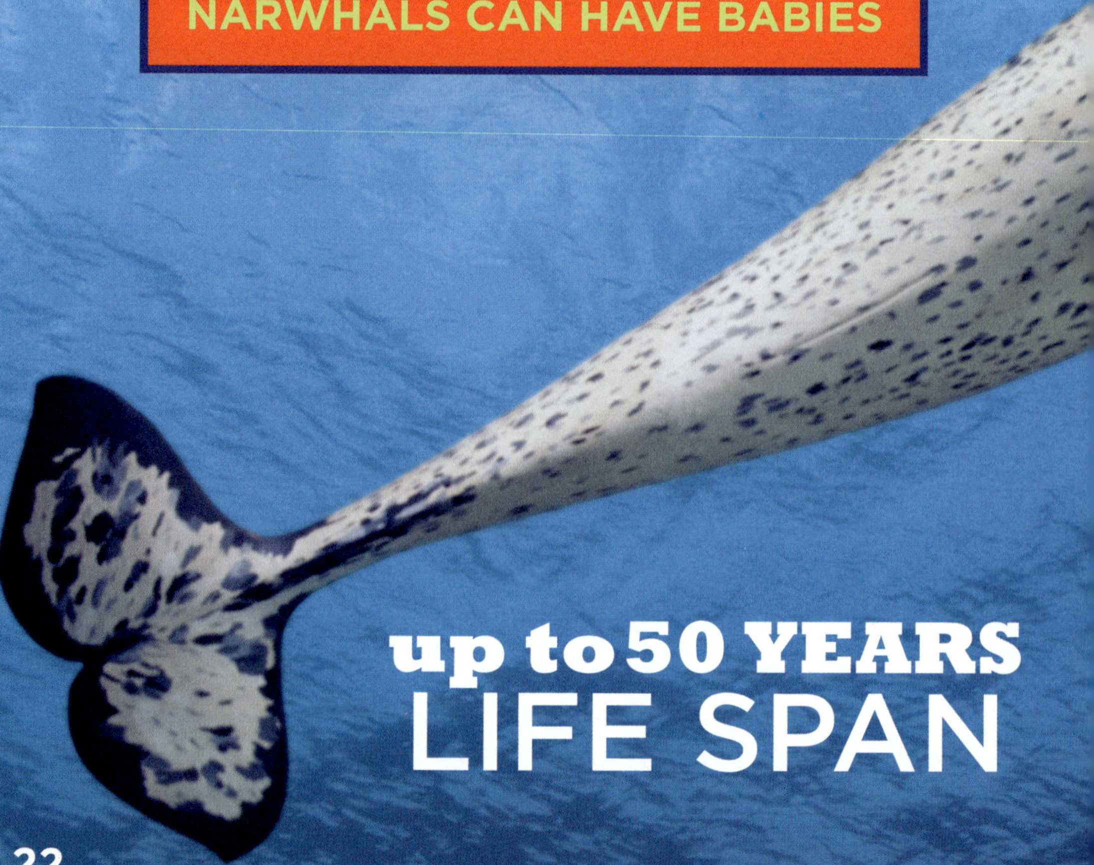

up to 50 YEARS
LIFE SPAN

up to
22 pounds
(10 kg)
weight of tusk

up to
9.8 FEET
(3 m)

LENGTH OF TUSK

more than
5,000 FEET (1,524 m)
how deep narwhals can dive

CHAPTER 4

DANGER!

Some large animals eat narwhals. Their main **predators** are polar bears and killer whales. Narwhals protect themselves, though. They hide from polar bears under the ice. They swim in water that's too deep or cold for killer whales.

Narwhal Food Chain

This food chain shows what eats narwhals. It also shows what narwhals eat.

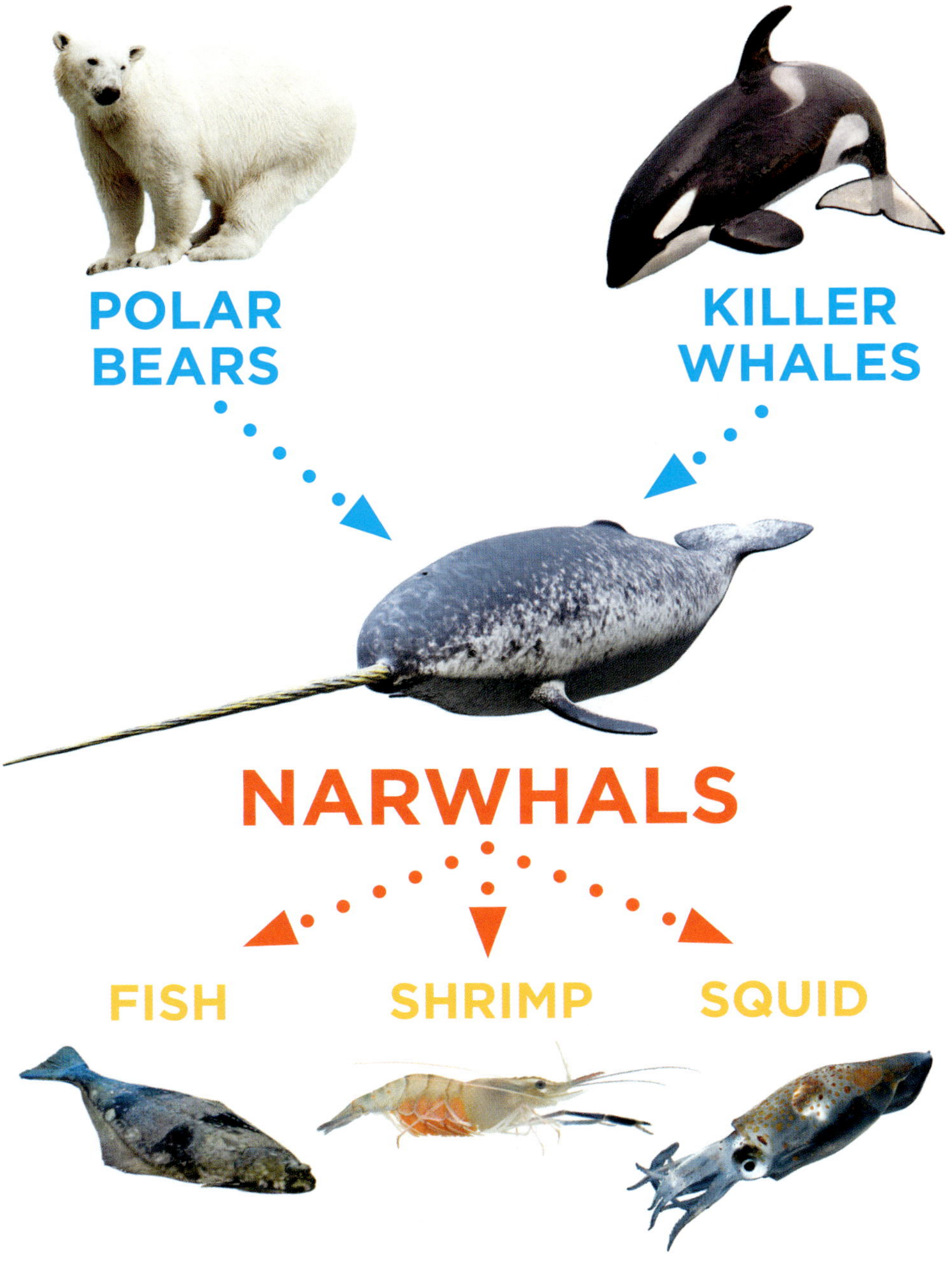

Warmer water also allows predators to move to new spots in the ocean. Killer whales have an easier time getting to narwhals.

Human Threats

Humans are narwhals' biggest threat. People hunt them. People also cause **climate change**. Climate change is a big problem for narwhals. Warmer temperatures melt sea ice. Less ice means narwhals have a harder time hiding from predators. Climate change also warms the water. Warmer water causes prey to move to new parts of the ocean. Narwhals have less food to eat.

Helpful Narwhals

Studying narwhals is difficult. They're shy. They swim places people can't go. But the narwhals that scientists do find can help them study climate change. Scientists have put sensors on these narwhals. The sensors measure ocean temperatures and how fast ice melts. Narwhals swim to places boats can't travel. The sensors send back valuable information. These creatures aren't just unique. They're helpful!

SENSORS

HELPFUL NARWHALS

SCIENTISTS

GLOSSARY

arctic (ARK-tik)—of or relating to the area of the North Pole or the region near it

blubber (BLUB-uhr)—the fat on whales and other large sea mammals

climate change (KLAHY-mit CHEYNJ)—changes in the Earth's weather patterns

herd (HURD)—a group of animals

mate (MAYT)—to join together to produce young; it also means one of a pair that joins together to produce young.

predator (PRED-uh-tuhr)—an animal that eats other animals

prey (PRAY)—an animal hunted or killed for food

social (SO-shul)—liking to be with and communicate with others

LEARN MORE

BOOKS

Holdren, Annie C. *Curious about Narwhals.* Curious about Wild Animals. Mankato, MN: Amicus, 2023.

Schuh, Mari. *Narwhals.* The World of Ocean Animals. Minneapolis: Jump!, Inc., 2022.

Szymanski, Jennifer. *Arctic Animals.* National Geographic Readers. Washington DC: National Geographic, 2023.

WEBSITES

Narwhal
oceanwide-expeditions.com/to-do/wildlife/narwhal-1

Narwhal Facts and Pictures
kids.nationalgeographic.com/animals/mammals/narwhal/

Unicorn of the Sea: Narwhal Facts
www.worldwildlife.org/stories/unicorn-of-the-sea-narwhal-facts

INDEX